7 Ways to Survive on Mars

Jill Bryant

Contents

Choose Adventure

Mars is the closest planet to Earth, but unlike Earth, it has an environment that is extreme and hostile for humans. The landscape is rocky, the soil is **toxic** and oxygen levels are very low. Yet, as far back as 1960, space agencies such as the USA's National Aeronautics and Space Administration (NASA) began sending devices to Mars, taking photos and collecting information. As we strive to learn more about Mars, some scientists and citizens are investigating the possibility of living on Mars one day.

Let's consider how humans could choose adventure and successfully live on another planet. What is required to survive on Mars?

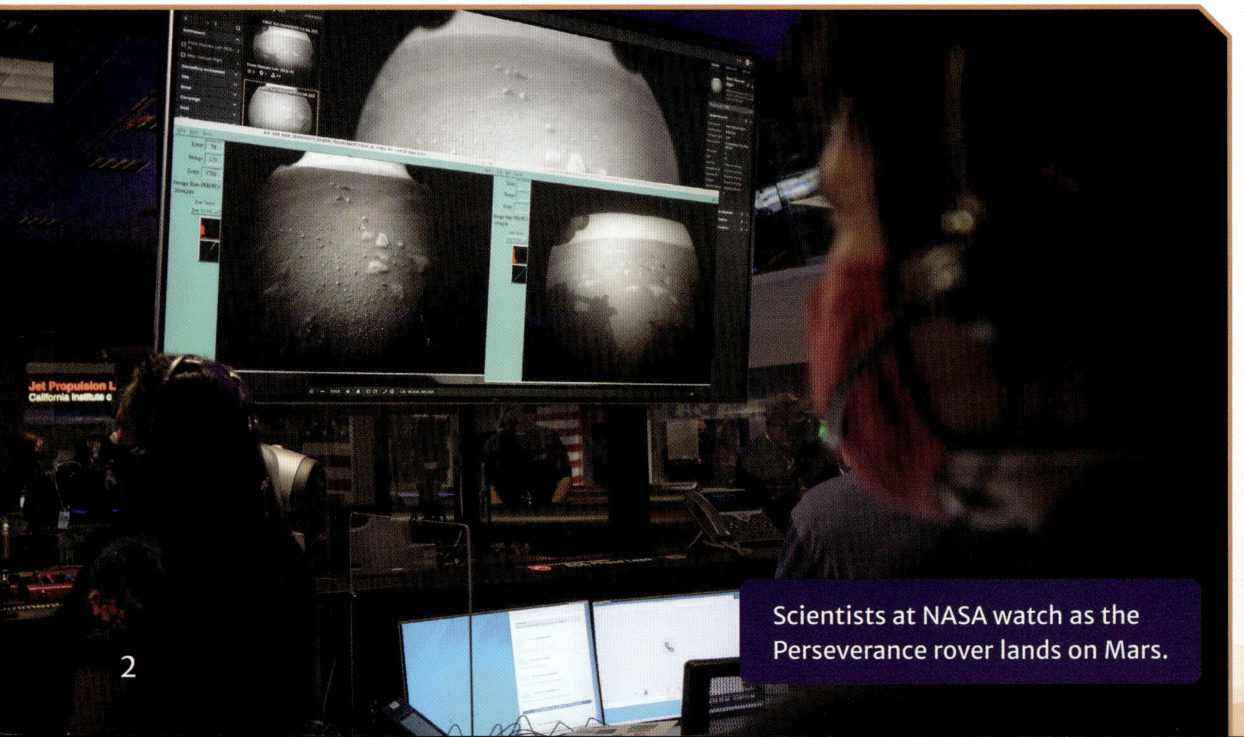

Scientists at NASA watch as the Perseverance rover lands on Mars.

Three and a half billion years ago, Earth and Mars weren't so different. But since then, Earth has become a planet that supports vibrant life, while Mars is cold and bare.

Earth and Mars: Then and Now

Earth – The Blue Planet		Mars – The Red Planet	
Then (3.5 Billion Years Ago)	**Now (Present Day)**	**Then (3.5 Billion Years Ago)**	**Now (Present Day)**
extreme heat	**temperate** climate	mostly cold, sometimes temperate	extreme cold
no oxygen	plenty of oxygen	no oxygen	trace amounts of oxygen
may have been covered in water	water (oceans, rivers, lakes) and ice	water (oceans, rivers, lakes)	may be some water; ice below surface
lots of active volcanoes releasing water vapour	hurricanes, tornadoes, volcanic eruptions, thunderstorms, snowstorms, droughts, floods	active volcanoes, floods, acid rain, dust storms, strong winds, superstorms, snow	dust storms
no living things	living things (flora and fauna)	possibility of living things	no confirmed living things

Ready, Set, Research!

To even consider living on Mars, you will first need to do lots of research.

Robotic devices called rovers have already collected lots of information for scientists on Earth. Space agencies such as NASA send rovers from Earth to Mars, a journey through space that takes seven to ten months. Then, the rovers travel over the Martian landscape, taking photographs. The rovers are "uncrewed", meaning there are no humans on board to operate the controls. Instead, commands are often sent from Earth.

The Perseverance rover has many cameras and robotic arms that allow it to take photos of Mars from many angles.

Rovers also take temperature readings, extract samples of rocks and soil, and search for tiny **organisms**. In the future, scientists hope to bring the sample tubes of rock and soil collected by each rover back to Earth to study. This process will be complicated and expensive but will allow scientists to gather even more detailed information about Mars.

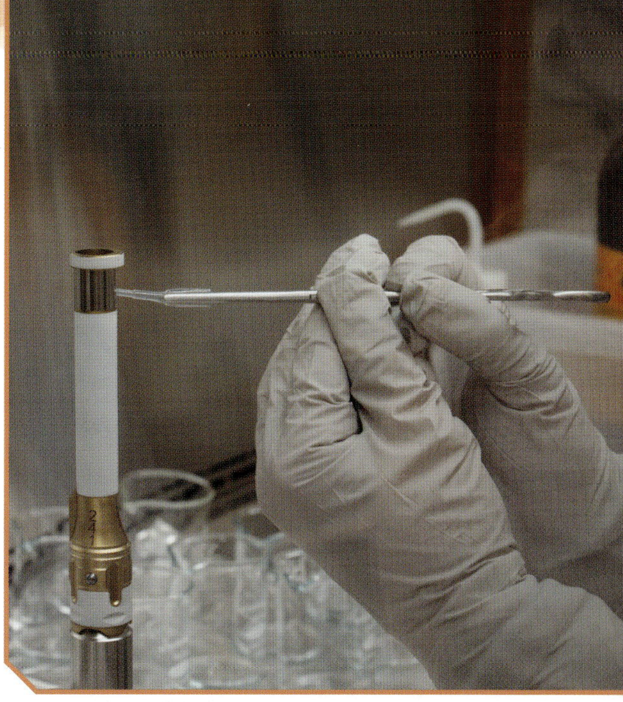

Samples of Martian rock and soil collected by rovers are stored in tubes like this.

When you have learned as much as possible about Mars, you and your team will be ready to start building a city.

NASA space engineers have sent five rovers to Mars.

- July 1997: Sojourner, a robot on wheels, hits the ground on Mars.
- January 2004: A rover called Spirit lands on Mars. Three weeks later its twin, Opportunity, arrives.
- August 2012: The Curiosity rover touches down for a long visit.
- February 2021: The Perseverance rover lands inside a Martian crater.

1 Make Your Own Oxygen

Humans need oxygen to survive. Unfortunately, the **atmosphere** on Mars has only small amounts of this gas. Mars's thin atmosphere is mostly made up of carbon dioxide, which is deadly to humans in large amounts.

To live on Mars, you will need a steady supply of oxygen.

The atmosphere of Earth is much thicker than that of Mars.

Oxygen Factory

Rather than bringing tanks of oxygen from Earth, you could make oxygen on Mars. It is possible to extract oxygen from the atmosphere on Mars with an oxygen generator.

In 2021, NASA used its Perseverance rover to launch a major experiment to produce a small amount of oxygen on Mars. It was successful! The experiment was named the **M**ars **Ox**ygen **I**n-Situ Resource Utilization **E**xperiment, or MOXIE.

MOXIE super-heats the small amounts of oxygen found in the atmosphere on Mars. Then, oxygen is extracted from the carbon dioxide in the atmosphere. Clean, pure oxygen is collected. Using this technology, scientists can attempt to create a stable supply of oxygen on Mars.

Astronauts need to breathe in about 30 grams of oxygen per hour of regular activity. However, MOXIE produces just 5 grams of oxygen per hour. Oxygen generators will need to become much more effective to support a city on Mars.

MOXIE was placed into the Perseverance rover in a lab in 2019.

Hold Onto Your Air

Living on Mars, you will be spending a lot of time indoors, in a controlled-atmosphere environment with air adjusted for humans.

To take a stroll on Mars, you will have to wear a spacesuit equipped with an air tank. As well as providing air, the suit will offer protection from harmful radiation – light emitted from the Sun in the form of waves or tiny particles.

Do not stay out too long, however. Running out of air has **catastrophic** results.

There is almost no oxygen on the surface of Mars.

A Spacesuit for the Future

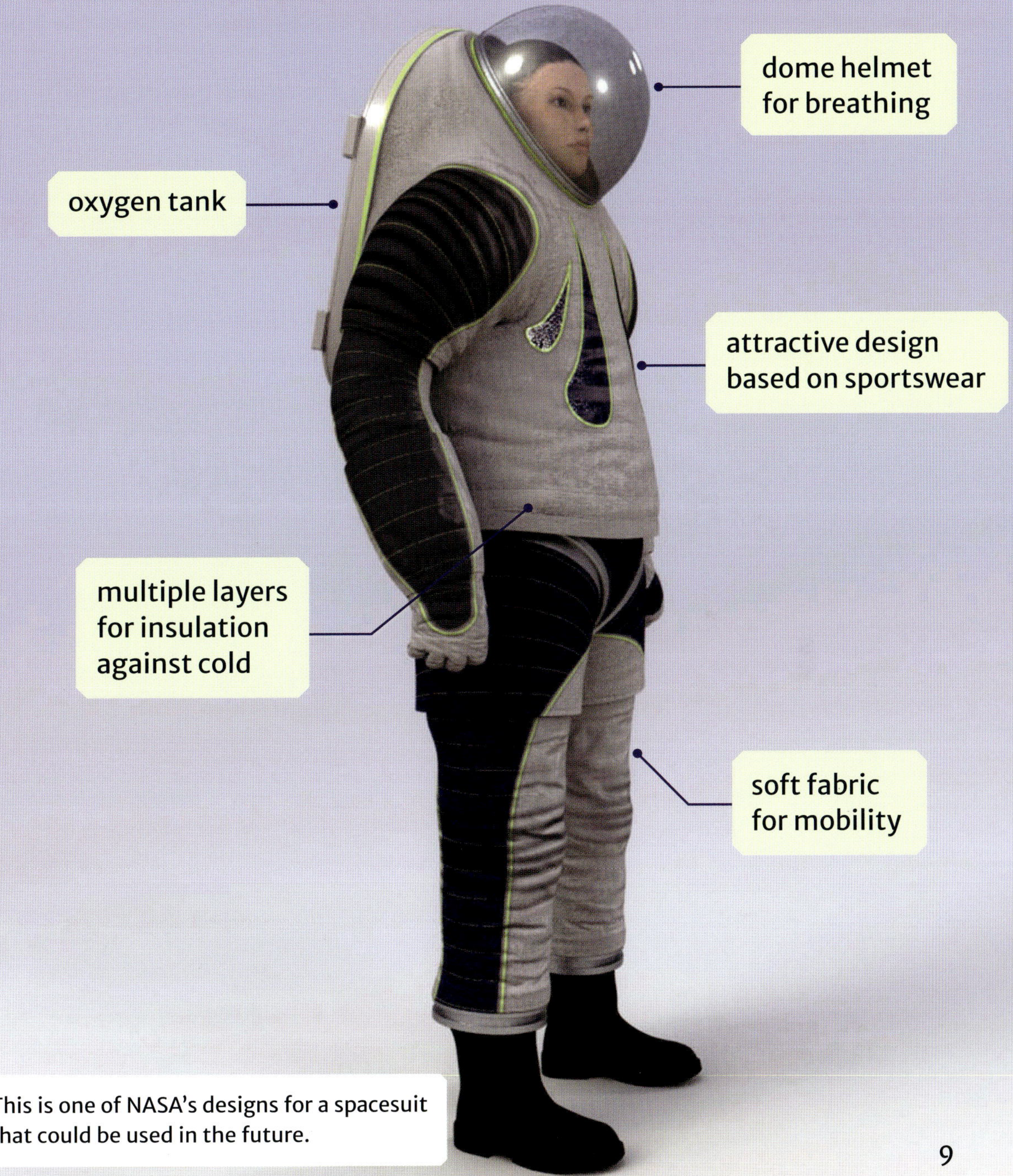

This is one of NASA's designs for a spacesuit that could be used in the future.

2 Extract Your Own Water

Water is essential for life, and a requirement for building a city on Mars.

Ancient Martian Brine

In 2000, scientists began studying images that suggested there might be water on Mars. It took more than 15 years to prove this was true. The discovery of ice below the Martian surface in 2016 changed everything. It showed that life on Mars could have existed, billions of years ago. This discovery also supports the possibility of life on Mars some time in the future. This is good news for Mars city planners.

The Korolev crater on Mars is filled with ice.

The solid water is not pure ice; it is a mixture of salt and ice called "brine". Brine was left behind when ancient Martian oceans and lakes dried up. Water can be extracted from the brine, providing a water supply on Mars.

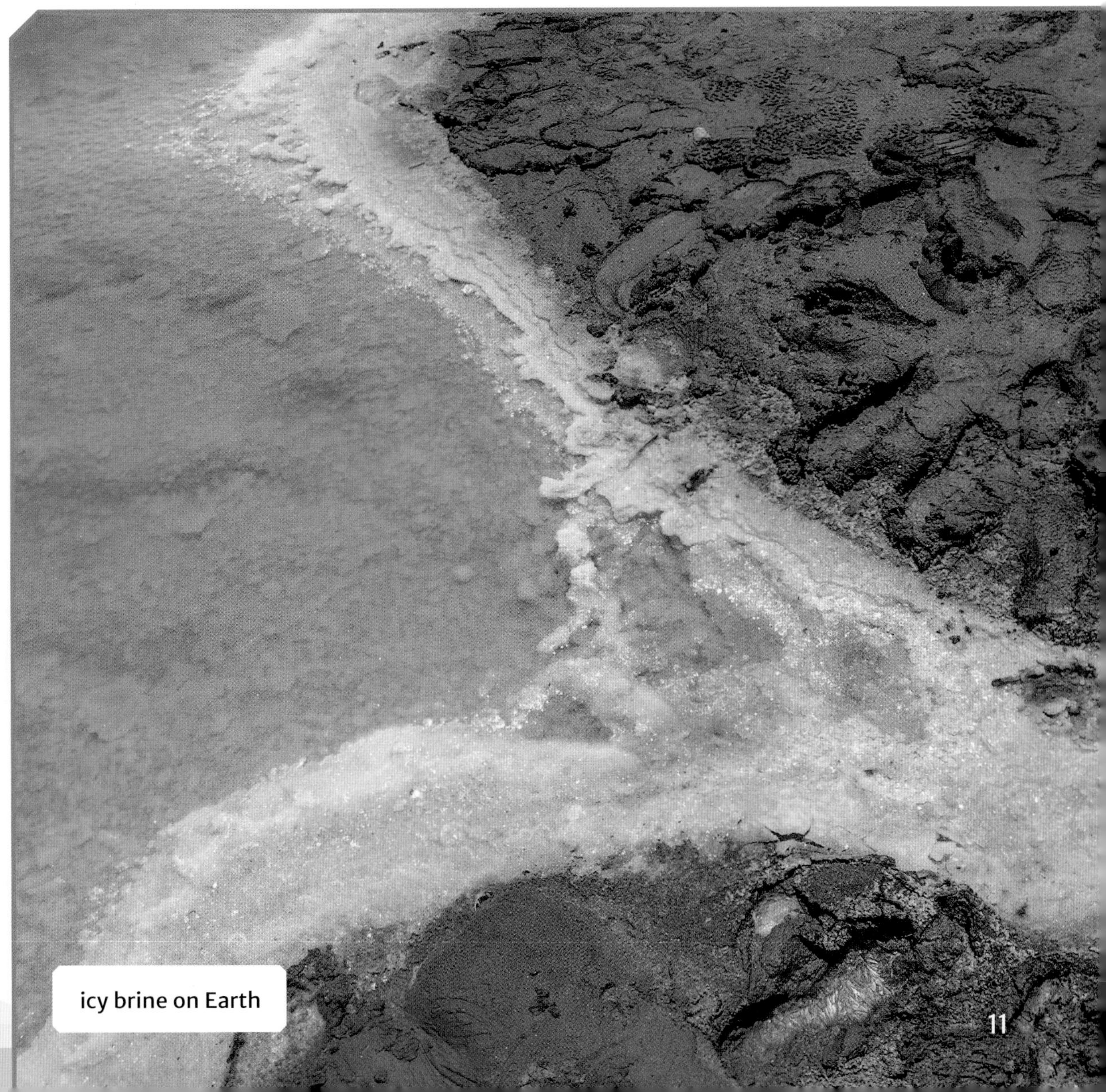
icy brine on Earth

Bottling It Up

Now you have to decide where to keep the precious water you have extracted. You will need a covered **reservoir** to store the city's water. Next, engineers will need to build an **infrastructure** with pipes that take water into homes and other buildings in the city. You will need to make sure the infrastructure is tough enough to withstand the severe dust storms on Mars.

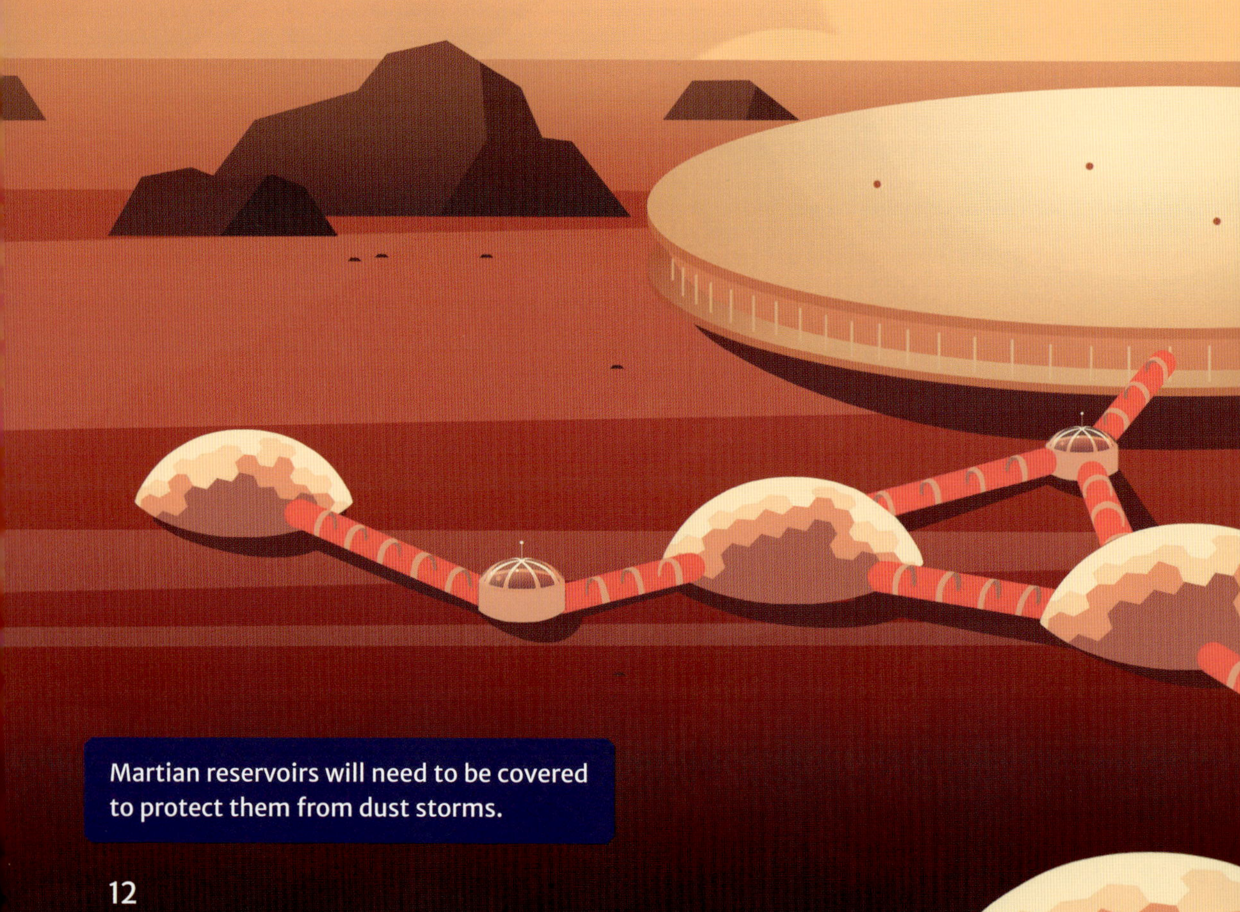

Martian reservoirs will need to be covered to protect them from dust storms.

Supply and Demand

To maintain a **sustainable** water supply, all people living on Mars will have to use water sparingly. Managed well, there will be enough water to drink and meet bathing and cleaning needs.

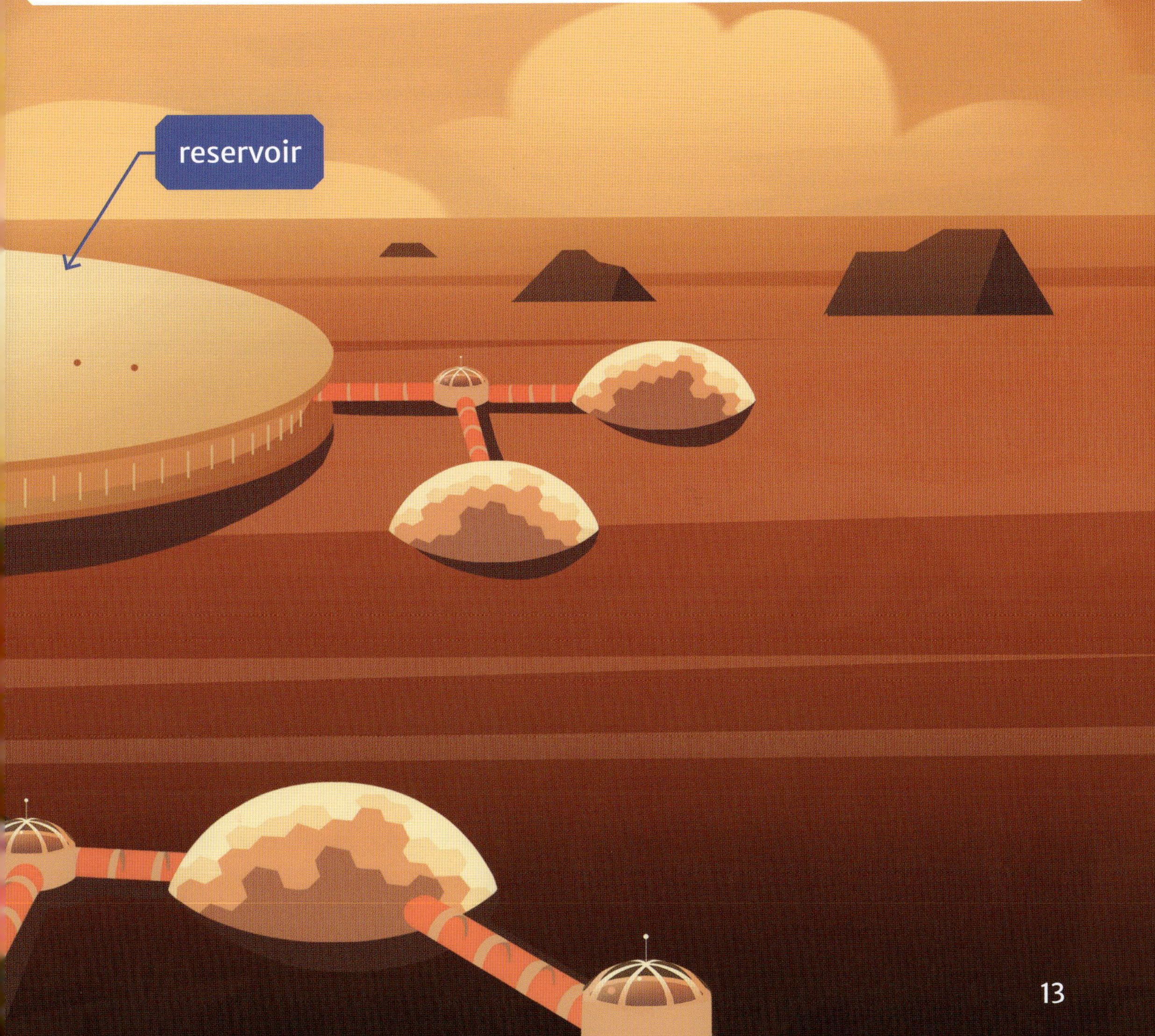

3 Grow Your Own Food

For years, astronauts on the **International Space Station (ISS)** have relied on food items such as orange drink crystals, soup in a tube, foil-wrapped dinners, dehydrated peaches and freeze-dried ice cream to meet their nutritional needs. For a mission to the Moon or the ISS, astronauts pack meals and snacks from Earth. But these Earth-made meal options are not sustainable for living for a long time on Mars.

Instead, you will need to grow your own food! You will need to build a controlled environment, such as a dome. Then you need seeds, soil, light and a watering system. This will allow you to enjoy a wide variety of nutritious foods.

To create a tasty menu for Mars, consider basic dietary requirements for humans: adequate calories, protein and essential nutrients.

An astronaut prepares to eat from a food packet on the ISS.

By testing crops grown indoors in the extreme conditions of Antarctica, food scientists can learn how to successfully grow crops on Mars.

Seeds

Seeds are small and will not be difficult to transport from Earth. Lettuce, kale, radishes, snow peas and chives are a few suggestions to try growing. Later, farmers can create a **seed bank** to ensure a steady supply.

Soil

Mars is covered in a crushed rock called "regolith", which is toxic for humans. To grow seeds in Martian soil, you will need to create better soil. By mixing in chemical elements that are found on Mars, it is possible to create a more fertile soil.

Experiments using volcanic soil from Hawaii, which is similar to Mars regolith, have led to good results. But so far, no one has grown anything in actual Mars regolith. It's possible that the first taste-testers will feel a little unwell from toxins that the plants could absorb through the soil.

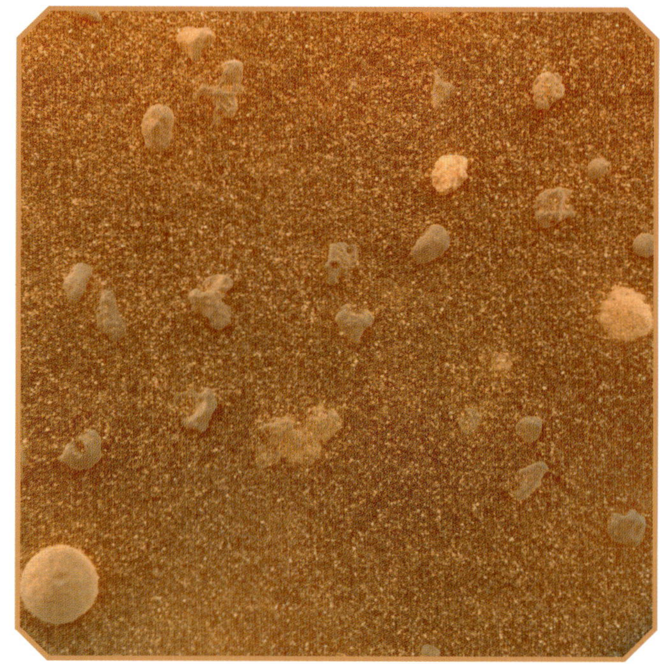

Martian regolith

Light Source

Plants need light to grow, but the extreme cold on Mars makes outdoor farming impractical. Instead, you will need to try indoor farming with "grow lights", gardening lights that mimic sunlight. Through experiments, scientists have found that plants that have been grown indoors – including wheat, potatoes and soybeans – thrive with a combination of red and blue **LED** grow lights.

leafy vegetables growing indoors under grow lights

4 Get a Roof over Your Head

Dust storms are not the only threat to consider when building a house on Mars. Some types of radiation can be deadly to living organisms, including humans. While Earth's atmosphere provides a natural protection against harmful forms of radiation, the thin Martian atmosphere does not. Therefore, homes on Mars will have to shield people from dangerous radiation.

Underground

One solution is to live underground, safely away from harmful rays. On Mars, lava flows from ancient volcanoes have left behind large underground passages called "lava tubes". These vast tunnels could become comfortable living spaces.

Lava tubes may be turned into underground homes on Mars.

On the Surface

For people who prefer natural light, living on or above the surface offers the best option. Windows in structures will allow sunlight to shine in. Or you may find a different way to allow light into living areas. Engineers will have to make sure harmful radiation does not seep through as well as light.

Shaping Up

Mars homes will look strikingly different from dwellings on Earth today. Shapes that mimic nature, such as domes, eggs or beehives, can withstand strong winds. The dwellings can be constructed using limited materials, which means fewer resources to take through space when heading to Mars.

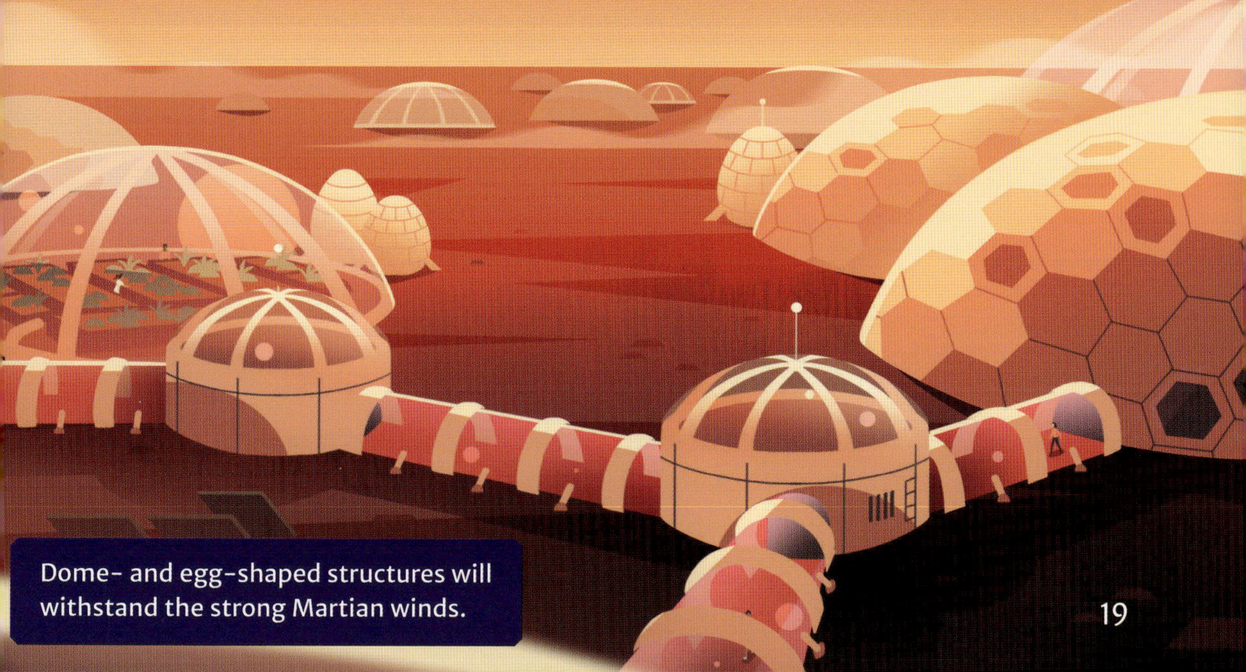

Dome- and egg-shaped structures will withstand the strong Martian winds.

What Building Materials Are Best?

Ice

After harvesting frozen water from below the Martian surface, robots can use 3D printing technology to print an ice structure in the shape of a dome. The structure will also need a protective **membrane**. This is to stop the ice from turning into vapour in the thin Martian atmosphere and blowing away.

Thick ice walls help protect people from radiation, but also allow light to seep through. Made from **translucent** materials, the dome ice structure will feature natural lighting during the day. For warmth, carbon dioxide could be added between the walls to act as insulation.

Ice structures on Mars will need a protective outer membrane.

Regolith

Experts recommend using 3D printing technology to help with construction on Mars. Some scientists think robots could build bricks out of the toxic Martian mineral regolith. Already, scientists have created 3D beams made from a Mars-like regolith. But this is new technology. You will need to test for toxins in the bricks, which could cause illness.

Martian regolith collected by the *Phoenix Mars Lander*

Basalt

Basalt is a hard rock found on Earth, the Moon and Mars. It is effective as a shield against radiation. Specially prepared basalt has proven to be five times stronger than concrete and is much lighter in weight than steel.

basalt on Earth

5 Exercise to Keep Healthy

Mars has a lower force of gravity than Earth, so on Mars you will weigh less and be able to jump higher! But beware: less gravity can lead to brittle bones, so regular exercise on Mars will be essential. Otherwise, your muscles will waste away and you will become very weak. Exercise that increases the heart rate, paired with weight training, will help to stop these effects.

Indoor Equipment

Indoor workout equipment might include treadmills and stationary bikes powered by solar energy. Other options are weight machines and free weights. You might also stock yoga mats, exercise balls and skipping ropes. Training sessions could include bouncing on a mini-trampoline.

Treadmills on Mars can be powered by solar energy.

Skipping will be a great way to exercise on Mars.

Extreme Outdoor Sports

Many features of Mars make it difficult to exercise outdoors there. The force of gravity is low, the atmosphere is thin, the temperature is extremely cold and the landscape is rugged. Still, if the lure of the outdoors is strong, the Martian landscape has a lot to offer. You may be interested in mountaineering, rock climbing or hiking.

To go on an outdoor excursion, there are rules you must follow. For travel on foot, you will have to wear a spacesuit. This will protect you from harmful radiation and the cold, and provide you with breathable air.

It is not safe to explore the Martian landscape alone. Be sure to find a friend or family member to go with you as you trek along rocky ridges and into the base of craters.

Residents of a Martian city may go out hiking together on the weekends.

6 Get Around with Big Wheels

To cruise expertly over the dangerous Martian terrain, you will need to use a specially designed rover. From your home, you will enter an indoor **pressurised** parking area and board a rover – no spacesuit required.

Then, you will drive through a climate-controlled tunnel with a series of sliding doors. Next, you will wait until the surrounding air in the tunnel adjusts to match the outdoor temperature and air pressure. Finally, you will exit onto Martian terrain. Inside the rover, you'll be protected from radiation, intense cold and strong winds.

Strong, tough wheels make your journey comfortable. The more wheels, the more stable your vehicle will be, so many rovers have more than four wheels.

Engineers at the Kennedy Space Center in the USA have designed a Martian rover for the future.

Driver's Checklist for Mars

Before you leave:

- Plan your route.
- Check the weather.
- Take a **positioning system** to help you find your way (if the area has been mapped!).
- Pack food and water.
- Inspect the rover's wheels.
- Check the rover's power supply.
- Test all rover instruments.

In transit:

- Take lots of pictures.
- Beware of craters, cliffs, boulders and other environmental dangers.
- Be prepared to return home at any time for safety reasons.
- Take care, if you choose to put on your spacesuit and exit the rover.

After you get back:

- Recharge your rover at a charging station.
- Record your experiences in the **logbook**.

7 Call Home Often

Regular contact with Earth may seem less important than securing air and food, but it is crucial to your **well-being**. You will have to figure out how to communicate with family and friends on Earth.

It is a huge challenge! The distance between Mars and Earth changes, depending on each planet's position in each of their orbits around the Sun.

Communicating by Laser

Laser communications is the ideal technology. It uses a kind of light that is invisible to humans. Earth's ground stations (with large satellite dishes) are positioned in clear-sky areas, so that clouds don't block the signal. With a state-of-the-art system, data will travel in seconds from Mars to a large satellite dish on Earth, where it will quickly be relayed to friends and family.

The Deep Space dish in Madrid, Spain, can communicate with rovers on Mars.

Let's Talk

For a city on Mars to succeed, it is critical that people have many opportunities to talk to friends and family back home on Earth. Sharing stories, laughing and connecting are important for morale and well-being. Good communication reduces feelings of loneliness. In the early days, when the population of Mars is very small, avoiding **social isolation** will be extra important.

The shortest distance between Earth and Mars is 54.6 million kilometres. The farthest distance between Earth and Mars is 401 million kilometres.

Calling home to friends and family on Earth will be very important.

Back to Earth

After several months of living on Mars, you may find yourself longing for home. Just remember – getting to Mars will be much easier than returning to Earth!

If you are determined to make the journey back to Earth, you will first need to visit the medical centre on Mars. Through a series of tests, you will find out if your bones are still strong enough to withstand both the journey back and the stronger force of gravity on Earth.

Then, you will need to collect fuel for your spacecraft. Hydrogen and oxygen are two important elements for rocket fuel. Stock up on vast stores of both. Eventually – with the help of technology, precision and science – you will be able to blast your rocket off Mars and head back to life on Earth as a citizen from Mars!

This is one possible design for a passenger spacecraft from Mars to Earth.

Glossary

atmosphere	the layer of gases surrounding a planet
catastrophic	causing huge damage
infrastructure	a set of structures and equipment that supports a service such as water supply
International Space Station (ISS)	a station for living and working on that orbits Earth, run by the space agencies of five countries
LED	Light Emitting Diode, a kind of light bulb that is very energy efficient
logbook	a written record added to over time
membrane	a very thin layer or barrier
organisms	living things, such as animals, plants or bacteria
positioning system	a system that uses signals from satellites to show a position on a map
pressurised	adjusted to have the same air pressure as on Earth
reservoir	a large lake used for human water supply
seed bank	a place where a collection of seeds from different kinds of plants is stored
social isolation	not having enough regular contact with other people

sustainable able to continue in the same way for a long time

temperate not extremely cold, hot, wet or dry

toxic poisonous to animals or plants

translucent allowing light to pass through

well-being the state of being comfortable, healthy or happy

Index